YOUR KNOWLEDGE HAS VALUE

T.S. Amar Anand Rao

Designing a Barophile Enrichment Apparatus to Culture Deep Sea Microbes

GRIN Verlag

Bibliografische Information der Deutschen Nationalbibliothek:

Die Deutsche Bibliothek verzeichnet diese Publikation in der Deutschen National-bibliografie; detaillierte bibliografische Daten sind im Internet über http://dnb.d-nb.de/ abrufbar.

Imprint:

Copyright © 2011 GRIN Verlag GmbH
Druck und Bindung: Books on Demand GmbH, Norderstedt Germany
ISBN: 978-3-656-09021-2

This book at GRIN:

http://www.grin.com/en/e-book/184271/designing-a-barophile-enrichment-appara-tus-to-culture-deep-sea-microbes

GRIN - Your knowledge has value

Der GRIN Verlag publiziert seit 1998 wissenschaftliche Arbeiten von Studenten, Hochschullehrern und anderen Akademikern als eBook und gedrucktes Buch. Die Verlagswebsite www.grin.com ist die ideale Plattform zur Veröffentlichung von Hausarbeiten, Abschlussarbeiten, wissenschaftlichen Aufsätzen, Dissertationen und Fachbüchern.

Visit us on the internet:

http://www.grin.com/

http://www.facebook.com/grincom

http://www.twitter.com/grin_com

Designing a Barophile Enrichment Apparatus to Culture Deep Sea Microbes

T.S. Amar Anand Rao

Molecular Biophysics Unit, Indian Institute of Science

Bangalore – 560012, Karnataka, India.

Abstract

Ever since the days of Holger W. Jannasch, the great deep sea microbiologist, the culturing and isolation of barophiles have come to existence. But using high pressure chemostat facilities are hard and high priced. We discuss a simple lab make barophile enrichment apparatus to culture and study deep sea microbes. We also isolate deep sea microbes that corrodes iron from sunken ships.

Keywords

Barophile, corrosion, chemostat, enrichment culture, barometer, one way valve

Introduction

Different microorganisms exist in different strata of the column and that some live in the aerobic and some in anaerobic zones. However, this is really where the discovery begins rather than ends! Explaining the complexity that lies within the depths of the ecosystem allows deeper insights into the microbial world (Rogan et al., 2005).

The Winogradsky column was developed and named after Sergei Winogradsky (1856-1953), a Russian microbiologist. He studied the complex interactions between environmental conditions and microbial activities using soil enrichment to isolate pure bacterial cultures (Madigan et al., 2000)

He studied the microbial organisms inhabiting sulphide-rich black mud ecosystems and pioneered our understanding of chemolithotrophy through his experiments with sulphate and nitrate reducing organisms(Tanner et al., 2000)

A continuous culture system that allows bacteria to be grown in steady-state populations under pressures of up to 700 atm (71 MPa) was constructed and tested. With readily available or slightly modified high-pressure chromatography equipment, a continuous flow of sterile medium is pressurized and passed through a 500-ml nylon-coated titanium reactor at flow rates of 0.01 to 10 ml min(sup-1). The pressure in the reactor is controlled by a backpressure regulator with greater than 1% accuracy. In test experiments, a culture of a psychro- and barophilic marine isolate from a depth of 4,900 m (strain F1-A, identified as a Shewanella sp.) was grown at 1, 300, and 450 atm (0.1, 30.4, and 40.5 MPa) and dilution rates of 60 and 90% of the organism's maximum growth rate (determined at 1 atm) in the required complex medium at levels of 3.3 and 0.33 mg of dissolved organic carbon per liter in the reservoir. Growth limitation by carbon was assured by an appropriate C/N/P ratio of the medium. The data indicate that barophilic growth characteristics in steady-state cultures of this psychro- and barophilic deep-sea isolate were positively affected by a decreasing growth rate at the higher of two substrate concentrations in the reservoir. After a 10-fold lowering of the substrate concentration, the effect was reversed. Under these conditions, the cell viability increased significantly, especially at the higher of the two pressures tested. The basic design of the system can principally also be used for growth studies on hyperthermophilic bacteria and archaea. (H W Jannasch, et al, 1996)

The influence of pressure on biological systems attracted the interest of scientists at Accademia del Cimento in Florence and of Robert Boyle in England during the seventeenth century. Science historian Stephen G. Brush credits Boyle for introducing 'a new dimension – pressure – into physics'. Euler provided the first mathematical definition of pressure in the eighteenth century. By the end of the nineteenth century, the concept of pressure had been developed into its present-day meaning. Pressure and temperature are today fundamental parameters for physical–chemical theory, for the description of environments, in industrial chemistry and biotechnology, and in both laboratory and ecological investigations of organisms. An indispensable aspect to the analysis of organisms inhabiting high-pressure environments is that temperature and pressure are coordinate variables. (A.A. Yayanos, Third Edition)

Barometers measure air pressure and are one of the most important instruments used in weather forecasting. Torricelli's barometer was the model for all mercury based barometers for the next two centuries, and providedan easy, if not always convenient way of measuringair pressure. (http://www.theweatherstore.com/ecbaforsc.html)

Corrosion of iron presents a serious economic problem. Whereas aerobic corrosion is a chemical process1, anaerobic corrosion is frequently linked to the activity of sulphate-reducing bacteria (SRB)2, 3, 4, 5, 6. SRB are supposed to act upon iron primarily by produced hydrogen sulphide as a corrosive agent3, 5, 7 and by consumption of 'cathodic hydrogen' formed on iron in contact with water2, 3, 4, 5, 6, 8. Among SRB, Desulfovibrio species—with their capacity to consume hydrogen effectively—are conventionally regarded as the main culprits of anaerobic corrosion2, 3, 4, 5, 6, 8, 9, 10; however, the underlying mechanisms are complex and insufficiently understood. Here we describe novel marine, corrosive types of SRB obtained via an isolation approach with metallic iron as the only electron donor. In particular, a Desulfobacterium-like isolate reduced sulphate with metallic iron much faster than conventional hydrogen-scavenging Desulfovibrio species, suggesting that the novel surface-attached cell type obtained electrons from metallic iron in a more direct manner than via free hydrogen. Similarly, a newly isolated Methanobacterium-like archaeon produced methane with iron faster than do known hydrogen-using methanogens, again suggesting a more direct access to electrons from iron than via hydrogen consumption.(Friedrich Widdel et al., 2004)

Materials and Methods

Procedure of making a Winogradsky column (Anderson et al 1999). The soil sample was cleaned of debris, stones, pebbles, grass clippings, leaves and moving insects.

Needed a narrow necked bottle bottle with a rubber cork stopper having two outlets or outlets.

A Winogradsky set up was designed (Picture 3) to create pressure more than the atmospheric pressure. Using one way walves to the outlet and inlet, the pressure inside the closed column was increased by closing the outlet and suppling air through pumping using the pipette bulb attached to the inlet.

A pipette rubber bulb was attached to a a short piece of aquarium tubing which is closed with a one way air valve purchased at a local aquarium shop. The outlet of this valve was further attached to a Millipore 0.45 micron syringe filter. The other end of this filter was inserted into one of the holes of the rubber stopper.

The other outlet of the rubber stop of winogradsky bottle was attached to a one way outlet air valve and closed with a open-close valve availabe in aquarium shops. The outlet of this valve attached to aquarium tubing was inverted a test tube funnel air collection appartus for analysis.

A Eco-celli Barometer purchased from (www.theweatherstore.com) was used to check the pressure of the air inside the bottle (Figure 2)

For this barometer to function, the outlet tube of the winogradsky pressure bottle was slided onto the open outlet of the Eco-celli barometer.

Or other option was to permanently attach the outlet of the Eco-celli barometer to a aquarium tube and insert it into the winogradsky bottle using a three holed rubber stopper.

An iron rod was also placed into the Winogradsky Pressure bottle to allow deep sea corrosive microbes to enrich.

Maintenace of constant pressure on the Pressure Winogradsky bottle was done by pumping in air through the bulb after closing the outlet of the bottle.

Results

Under constant maintenance of pressure a little more than 1 atm rust and black colour biofilms patterns were dominant indicating their ability to grow in the depths of the water body (Figure 1).

Temperature was maintained by placing the setup in algar growth chambers with air coolers to simulate the cold of the deep sea temperatures.

With time, black anaerobic zones of microbe belonging to sulfate-reducing bacteria formed in the mud region. *Desulphovibrio* sp. reacts with iron to produce black ferrous sulphide.

Most of the water was covered from rust of the iron. Also, the photosynthetic bacteria coloured bright red by a large population of purple non-sulphur bacteria including species of *Rhodopseudomonas*, *Rhodospirillum* and *Rhodomicrobium*.

Figure 1

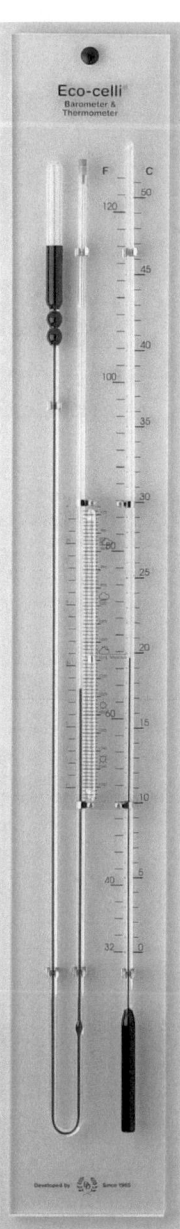

Figure 2

One-Way valve

Open-Close Valve

2 holed Rubber-stop

Millipore
Filter
0.45u

Pipette
Bulb

Inverted Test tube gas collection

Figure 3

Discussion

The pressure can be increased to more than 1 atm if all the components of the pressure winogradsky bottle including the tubings are made to resist the pressure by using high pressure resistant PVC plastic materials.These materials are used in submersiles that go deep oceans. And also this setup can be furnished accordingly and modified in future. Thus this is a tool to simulate deep sea pressure environments and thus enrich the microbes of barophilic environments and also suggest how deep sea corrosion of sunken ship iron by chemolithotrophy microbe interactions. Using these tool, further pressure temperature gradient correlation studies on barophilic microbes can be done for understanding better.

References

1. Anderson, Delia C, and Hairston, Hosalina V, The Winogradsky column and biofilms, models for teaching nutrient cycling and succession in an ecosystem, The American Biology Teacher, Vol. 61, No. 6, June 1999, pp. 453-459

2. Rogan, B., M. J. Lemke and M. Levandowsky. 2005. Exploring the sulfur nutrient cycle using the Winogradsky column. American Biology Teacher. 67:279-287

3. Madigan, M. T., Martinko, J. M. & Parker, J. (2000). Brock Biology of Microorganisms, 9th Edition. Upper Saddle River, NJ: Prentice-Hall.

4. H W Jannasch, C O Wirsen and K W Doherty, A pressurized chemostat for the study of marine barophilic and oligotrophic bacteria. Appl. Environ. Microbiol. May 1996 vol. 62 no. 5 1593-1596

5. A.A. Yayanosa, High-Pressure Habitats, Encyclopedia of Microbiology (Third Edition), Pages 228-239

6. David R. Drake1 and Kim A. Brogden2 Bookshelf, U.S. National Library of Medicine, National Institutes of Health,Bookshelf ID: NBK2488, Chapter 2Continuous-Culture Chemostat Systems and Flowcells as Methods to Investigate Microbial Interactions

7. http://www.theweatherstore.com/ecbaforsc.html

8. Hang T. Dinh, Jan Kuever, Marc Mußmann, Achim W. Hassel, Martin Stratmann, Friedrich Widdel, Iron corrosion by novel anaerobic microorganisms, Nature 427, 829-832 (26 February 2004)